I0482046

HOW TO MANUALLY UPDATE YOUR KINDLE

The Ultimate Guide for Complete Beginners On How to Manually Update Your Kindle in Few Minutes.

BY

CHARLES S. MILLS

COPYRIGHT

Charles S. Mills

TABLE OF CONTENT

CHAPTER 1

INTRODUCTION

The Kindle device is one of the world most popular android tablet with over 3 million people who owns it, and enjoy using the amazing features it has in it.

Apart from the popular use of kindle device as an E-reader, it can also perform other good and interesting functions. The user interface is a friendly one and it has a lot of good package in it.

This guide will show you how you can manually update your Kindle with step by step process even as a beginner, all you have to do is just follow the steps as instructed. But if you find this very difficult, there is no needs to worry because I got you cover.

All over the world millions of people haven't been able to use and manually update kindle but dis book gives the breakdown of all solution to any problem you might encounter. With proper medication and better understanding you will update your kindle manually in a few minutes

Thankfully each steps are very easy and simple to follow, that even a beginner can master it in a few minutes.

CHAPTER 2

HOW TO MANUALLY UPDATE YOUR KINDLE

If you wish to have the latest Kindle features so quick, the suitable way to get an immediate update for your Kindle is doing it manually. Below you will be shown how you can easily update your Kindle.

As a general rule, Amazon's over-the-air updates are most often hiccup free (nevertheless they could take upwards of a month to slowly roll out to every Kindle globally). But in any ways, maybe your Kindle stopped updating for no reason (like ours did), or you could be in a haste to get the latest and very good features. However, whatever reasons you have, you don't need to wait for Amazon to drop that update out.

CHAPTER 3

IDENTIFY YOUR KINDLE MODEL

Notwithstanding a second generation Kindle Paperwhite updating will be done in this tutorial, the same skills you will be guided through your works on all the different Kindle models. The first and foremost step to take is identifying which Kindle you have, ensuring you are comparing the correct software version for your model and downloading the suitable update.

Rather than skelly at the tiny model number on the back of your case and then Google the model number, the easiest way is to inspect the serial number as the first four alphanumeric characters show the model or generation of your Kindle.

If your Kindle is connected to your Amazon account, just log into your Amazon Content and Devices dashboard and then click on the "Your Devices" tab. After that you should choose the Kindle you wish to update and read the notation next to it, like so:

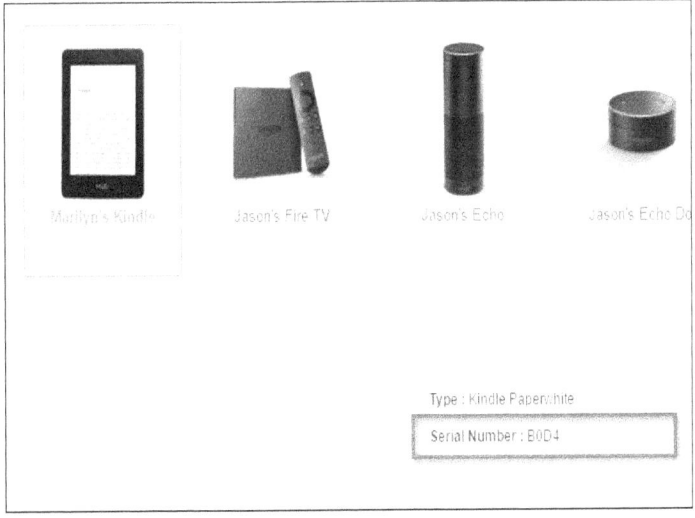

In the case of the Kindle am interested in updating now, the first four character of the serial number are B0DA. Although you can also find the serial number by switching on your Kindle and looking in the;

Setting

Device info menu.

Once you've gotten the first four characters, the following list can be reference to figure out exactly which model you have. Making it easy you can use Ctrl+F to search for the character string.

a. Kindle 1 (2007): B000

b. Kindle 2 (2009): B002, B003

c. Kindle DX (2010): B004, B005, B009

d. Kindle Keyboard (2010): B006, B008, B00A

e. Kindle 4 (2011): B00E, B023, 9023

f. Kindle Touch (2012): B00F, B010, B011, B012

g. Kindle Paperwhite 1 (2012): B024, B01B, B01C, B01D, B01F, B020

h. Kindle Paperwhite 2 (2013): B0D4, 90D4, B0D5, 90D5, B0D6, 90D6, B0D7, 90D7, B0D8, 90D8, B0F2, 90F2, B017, 9017, B060, 9060, B062, 9062, B05F, 905F

i. Kindle 7 (2014): B001, B0C6, 90C6, B0DD, 90DD

j. Kindle Voyage (2014): B001, B013, B053, B054

k. Kindle Paperwhite 3 (2015): G090

l. Kindle Oasis (2016): G0B0

m. Kindle 8 (2016): B018

Now after you have checked your serial number twice against the list, it's time to hold the real update files.

CHAPTER 4

GET THE UPDATE DOWNLOADED

Equipped with the version number of your Kindle in our case, verified by the serial number, the Paperwhite 2 head go to the Amazon Fire and Kindle Software Update page. Navigate down until you arrive at the Kindle section and then you should match up the Kindle you have with the fitting model. Keep in mind that there might be several versions of the same model, this indicate the reason we went searching for the serial number in the step above.

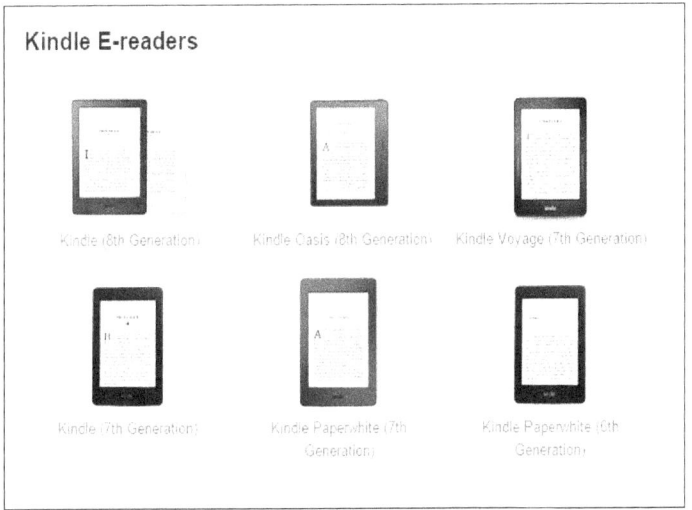

Kindle E-readers

Kindle (8th Generation) Kindle Oasis (8th Generation) Kindle Voyage (7th Generation)

Kindle (7th Generation) Kindle Paperwhite (7th Generation) Kindle Paperwhite (6th Generation)

Immediately after you have chosen the model, you will see a download link with the present version number listed. Pay attention of the version number but don't download it yet.

Download Kindle Paperwhite Software Updates

Determine the current software version on your Kindle Paperwhite before downloading and installing a software update. For more information, go to Determine Your Kindle Paperwhite (6th Generation) Software Version.

Click the link below to download the software update to your computer.

Download Software Update 5.8.5.0.2

In accordance with certain free and open source software licenses, Amazon is pleased to offer an archive file of source code for this software update.

Before downloading the update, verify that version number is higher than the present version on your Kindle. On your Kindle, head to;

Menu

Settings

Menu

Device info.

You will see a pop up screen like the following one.

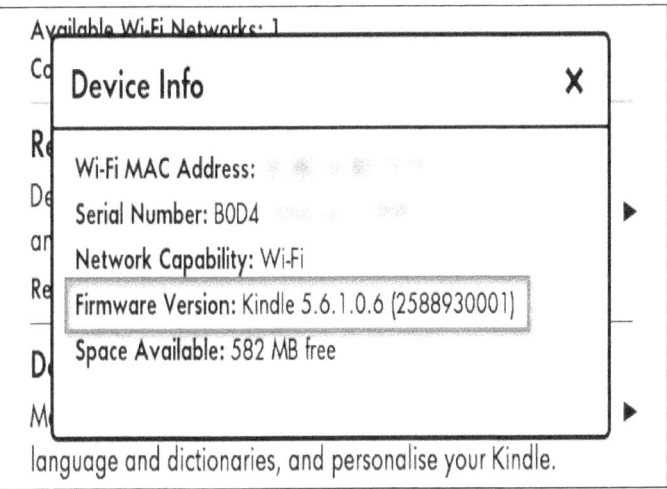

Obviously, the firmware version we have on our Paperwhite (5.6.1) is in back of the most current version as of this writing (5.8.5). somewhere along the line, we obtained the summer 2016 update but missed out on the big fall 2016 update that brought in the new home screen plan. Now, with the difference between the most recent version and our version assured, we are able to download the update file. All you have to do is just click on the link "Download Software Update (version number)". Then the update will be save as a bin file.

CHAPTER 5

GET UPDATE COPY TO YOUR KINDLE AND INSTALL IT

As soon as the download is done, connect your Kindle Paperwhite to your computer using a USB cable and copy the update BIN file to the root directory of your Kindle Paperwhite. Ensure to put the file in the top level folder, so if your computer mounts the Kindle as the F drive, after that ensure that the part to the copied

update package should be
F:/update_kindle_ (version number).bin

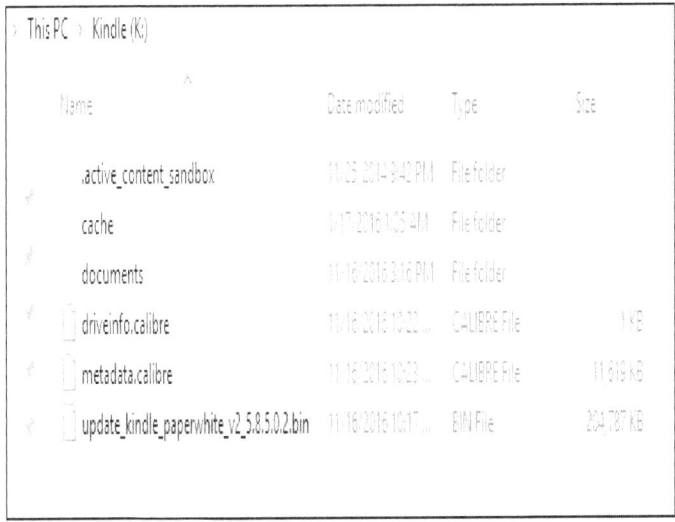

As soon as the file is placed in the root directory of your Kindle device, you should right click on the Kindle's drive and select Eject to unmount it from your computer. You can unplug it.

Now, you just need to order the Kindle to update via the Kindle's menu system. On the Kindle, head to the menu > settings to have an access to the settings menu, then you should click on the menu button again within the settings menu and choose "Update Your Kindle". Now, press OK and wait. (if the "Update Your Kindle" option is grayed out, indicating that the Kindle wasn't able to locate the .bin file; be assured you placed you put it in the root directory and try again.

After rebooting your Kindle (you shouldn't be concerned if it takes some minutes or longer to complete rebooting and updating), run the version check method by looking in the device info menu. There, you should see an updated version number, and with latest Kindle operating techniques releases, you are able to read the release notes right on your device by clicking the "More Info" button.

Your Kindle is now updated with the newer features and without waiting for the device to update over the air automatically. With belief that your future

OTA updates are smoother, its easy (knowing where to look at) to catch an update manually and reload your Kindle to the recent version.

THE END

You can check out my other books by the same publisher:

HOW DO I INSTALL GOOGLE PLAY ON KINDLE FIRE

HOW TO SETUP A KINDLE FIRE HD

AMAZON ECHO SHOW

AMAZON ECHO DOT

HOW DO I RESET MY KINDLE

HOW TO DELETE BOOKS IN KINDLE

Charles S. Mills

HOW DO I SIDELOAD APPS INTO MY KINDLE
FIRE

HOW TO SETUP YOUR AMAZON ECHO

HOW TO UNLOCK AMAZON FIRE STICK

HOW TO SETUP YOUR NEW CHROMECAST

HOW TO TRANSFER BOOKS TO KINDLE,
KINDLE FIRE, AND KINDLE APP

Charles S. Mills

Charles S. Mills

www.ingramcontent.com/pod-product-compliance
Lightning Source LLC
Chambersburg PA
CBHW072034230526
45468CB00021B/1788